Sharing the Journey
A Military Spouse Perspective

Dawn A. Goldfein
Spouse of the 21st Chief of Staff
Gen David L. Goldfein, USAF

with

Dr. Paul J. Springer

and

Capt Katelynne R. Baier, USAF

Air University Press
Maxwell Air Force Base, Alabama

Chief of Staff, US Air Force
Gen David L. Goldfein

Commander, Air Education and Training Command
Lt Gen Marshall B. Webb

Commander and President, Air University
Gen James B. Hecker

Director, Air University Press
Maj Richard Harrison

Project Editor
Dr. Stephanie Havron Rollins

Illustrators
Timothy Thomas
L. Susan Fair

Print Specialist
Megan N. Hoehn

Air University Press
600 Chennault Circle, Building 1405
Maxwell AFB, AL 36112-6010

https://www.airuniversity.af.edu/AUPress/
Facebook: https://www.facebook.com/AirUnivPress
and
Twitter: https://twitter.com/aupress

AIR UNIVERSITY PRESS

Library of Congress Cataloging-in-Publication Data

Names: Goldfein, Dawn A., author. | Springer, Paul J., author. | Baier, Katelynne R., author. | Air University (U.S.). Press, issuing body.
Title: Sharing the journey : a military spouse perspective / by Dawn A. Goldfein, with Paul J. Springer and Katelynne R. Baier.
Other titles: Military spouse perspective
Description: Maxwell Air Force Base, Alabama : Air University Press, 2020. | "Published by Air University Press in July 2020"—CIP data page. | Includes bibliographical references and index. | Summary: "A unit's command team is the partnership among the commander, the senior noncommissioned officer (NCO), and a volunteer lead spouse. As the primary advisor, ambassador, and advocate for the spouses and families of members in the unit, finding the right person to undertake the important role of volunteer lead spouse is one of the most important decisions a commander will make. Once a spouse in the unit decides to take on the role, it can be challenging and incredibly rewarding to navigate working with military leadership, state or local government, base programs and organizations, and other military spouses to take care of families. This book captures "words of wisdom" collected by Mrs. Dawn Goldfein, spouse of the 21st Chief of Staff of the Air Force, Gen David L. Goldfein over their 37-year career. For command teams that seek to understand and leverage the military "spouse network" of command, lead, key, and key spouse mentors within their unit or their installation, it offers a treasure trove of useful ideas and stories"— Provided by publisher.
Identifiers: LCCN 2020033022 (print) | LCCN 2020033023 (ebook) | ISBN 9781585663156 (paperback)
Subjects: LCSH: United States. Air Force—Military life. | Air Force spouses—United States. | Social role—United States. | Leadership—United States. | Families of military personnel—United States.
Classification: LCC UG633 .G6135 2020 (print) | LCC UG633 (ebook) | DDC 358.4/1120973—dc23 | SUDOC D 301.26/6:SP 6
LC record available at https://lccn.loc.gov/2020033022
LC ebook record available at https://lccn.loc.gov/2020033023

Published by Air University Press in July 2020

Disclaimer

Opinions, conclusions, and recommendations expressed or implied within are solely those of the authors and do not necessarily represent the official policy or position of the organizations with which they are associated or the views of the Air University Press, Air University, United States Air Force, Department of Defense, or any other US government agency. This publication is cleared for public release and unlimited distribution.

Reproduction and printing are subject to the Copyright Act of 1976 and applicable treaties of the United States. This document and trademark(s) contained herein are protected by law. This publication is provided for non- commercial use only. The author has granted a nonexclusive royalty-free license for distribution to Air University Press and retains all other rights granted under 17 U.S.C. §106. Any reproduction of this document requires the permission of the author.

This book and other Air University Press publications are available electronically at the AU Press website: https://www.airuniversity.af.edu/AUPress.

This book is dedicated to my mentor, Nancy Diehl, the first squadron lead spouse who inspired me to lead other spouses throughout our 37-year Air Force journey.

Contents

Dedication	*iii*
Foreword	*vii*
About the Authors	*ix*
Preface	*xi*
Acknowledgments	*xiii*
Abstract	*xv*
1 The Volunteer Lead Spouse: Qualities, Roles, and Definitions	1
Command Spouse	3
Lead Spouse	3
Key Spouse	3
Key Spouse Mentor	3
2 Words of Wisdom for the New Lead Spouse	7
3 Transitions: How to Step In and Out of Command	11
Where to Start: Stepping into Command	11
Setting Your Successor up for Success	14
4 Leveraging the Spouse Network	17
5 Common Challenges and Questions	21
Resources	35
Conclusion	41
Abbreviations	45
Bibliography	47
Index	49

Foreword

Of all the blessings serving as the 21st Chief of Staff of the Air Force (CSAF), by far the best is the opportunity to do the job every day side by side with my best friend, high school sweetheart, the love of my life, and the 21st First Lady of the Air Force, Dawn. She wrote this book, *Sharing the Journey,* as a companion to my work, *Sharing Success—Owning Failure.*[1] It is written for the entire command team to help leaders understand the critical role of a volunteer lead spouse, an essential member of every unit. I don't believe there is a parallel work of its kind, and it is long overdue. In these pages, we hope to capture the vital role lead spouses play in a unit. It is filled with stories collected during 37 years of marriage and throughout our Air Force journey together.

As military leaders, our volunteer lead spouses stand with us to accomplish the most sacred duty we have as commanders: taking care of our Airmen and their families. We expect Airmen to be ready to deploy at a moment's notice to support the mission. In return, they trust us to support and protect their families while they are gone. We have been a busy Air Force over the past two decades, focused primarily on wars in Iraq and Afghanistan. Today, as we continue our focus on revitalizing the fundamental fighting formation of the United States Air Force (USAF)—the squadron—we rely on our command teams more than ever to care for our Airmen. The most successful squadron commanders understand and embrace their responsibility for building a culture of trust and cohesion across their unit. Who is better equipped to help create a vibrant, caring community in the unit than our gifted and capable Air Force spouses? This only happens if commanders allow and empower our spouses to take the initiative.

Our military spouses exhibit a special kind of courage when enduring long hours, separations, hardships, and often loneliness that accompany military duty in a nation at war. Like so many spouses, Dawn has adjusted her career and dreams to support mine. She moved 21 times and sent me off to war four times while raising our two incredible daughters, often alone. She also experienced every combat pilot spouse's worst nightmare when the wing commander and chaplain arrived at her door in the middle of the night to inform her that her husband was shot down in enemy territory and his condition was unknown.

Like so many other volunteer lead spouses throughout our history, Dawn has been an incredible role model for others to follow. Not only do I love her dearly, but I also admire her strength, poise, and passion for Air Force families. *Sharing the Journey* is the perfect companion to *Sharing Success—Owning*

Failure. Dawn partners with other senior spouses to offer experiences, wisdom, and perspective to answer the question she and several other lead spouses have asked me since *Sharing Success—Owning Failure* was published, "Where is my chapter?" *Sharing the Journey* is a must-read for every command team.

As we complete our tour of duty as your 21st CSAF and 21st First Lady of the Air Force, Dawn and I depart truly honored to have served with each of you. We will be cheering on all who follow. America sleeps well at night because our Airmen do not. You stand the watch as our nation's sentinels. It has been our honor to have stood beside you.

Fight's on!

DAVID L. GOLDFEIN
General, USAF
21st Chief of Staff
Proud Husband of the 21st First Lady of the Air Force

Notes

1 David L. Goldfein. *Sharing Success—Owning Failure: Preparing to Command in the Twenty-First Century Air Force (*Maxwell Air Force Base, AL: Air University Press, 1999), https://media.defense.gov/.

About the Authors

Mrs. Dawn A. Goldfein is married to Gen David L. Goldfein, the 21st Chief of Staff of the USAF. She claims Texas as her home, although she grew up in a military family. Continuing the family tradition of a military lifestyle, she has accompanied her husband on 19 assignments around the globe, and they are now completing their latest adventure in Washington, DC. Inspired by a call to serve, she has dedicated her life to supporting her family and the Airmen and families of the USAF.

Mrs. Goldfein graduated with honors from Midwestern State University and holds a bachelor of science degree in education with a minor in biology. She has taught kindergarten, first grade, and third grade over the course of 29 years.

Mrs. Goldfein is a full-time volunteer supporting and advocating on behalf of military families and children. She is an avid supporter of military families through her service in many organizations, such as the Military Child Education Coalition (MCEC), the Ambassador Committee for Easter Seals, the Tragedy Assistance Program for Survivors (TAPS), United Through Reading, and ThanksUSA, to name only a few. Additionally, she represents the Air Force as an Arlington Lady. For her countless hours of service, Mrs. Goldfein was recognized with the Distinguished Public Service Award in 2015.

General and Mrs. Goldfein met in high school in Kaiserslautern, Germany, during the summer of 1976. They were married in Colorado Springs the day after Dave's graduation from the Air Force Academy (didn't want to rush) and have enjoyed 37 years together on their Air Force journey. They have two incredibly talented daughters who each followed them into their chosen profession. Danielle is a third-generation officer serving in the USAF; and Diana is a third-generation school teacher and fitness instructor. Both girls married their Texas A&M boyfriends who are wonderful young men that the Goldfeins are proud to call their sons-in-law. Two years ago, Dave and Dawn earned their most important callsigns ever as "Papa and Nana" when they were blessed with two gorgeous granddaughters.

Dr. Paul J. Springer is a professor of comparative military studies and the chair of the Department of Research at the Air Command and Staff College. He holds a PhD in history from Texas A&M University, an MA in history from the University of Northern Iowa, and a BS in psychology from Texas A&M University. Prior to joining the ACSC faculty, he taught at the US Military Academy at West Point and Texas A&M University. He teaches courses on war theory, leadership, airpower, strategy, military history, military command, and terrorism. He is the author or editor of 12 books in print, with another four in progress and scheduled for publication in 2021. His most prominent works include *America's Captives: Treatment of POWs from the Revolutionary War to the War on Terror* and *Outsourcing War to Machines: The Military Robotic Revolution*. He is a senior fellow of the Foreign Policy Research Institute and the series editor for the History of Military Aviation and Transforming War series, both with the US Naval Institute Press.

Capt Katelynne R. Baier is the aide-de-camp to the Chief of Staff of the Air Force. As aide-de-camp, she acts as the CSAF's personal assistant and trusted advisor, responsible for domestic and international travel, conferences, and event execution.

Captain Baier graduated from Trinity University with a double major in international relations and Chinese language. She received her commission from the Reserve Officer Training Corps Detachment 842, University of Texas–San Antonio. She is a graduate of the United States Weapons School.

Preface

Congratulations on entering a new phase of your Air Force journey. Being a member of a command team will likely be the most challenging and most rewarding role you will experience as a part of military life. The intent of this publication is to pass on "words of wisdom" to help the entire command team prepare for the task ahead by describing the role of the volunteer lead spouse and all the ways they may choose to serve. This is not a "to-do list" nor will it answer all of the questions about "how to be a successful volunteer lead spouse;" rather, even though our Air Force and Space Force are changing every day, this publication captures timeless values, principles, and ideas to consider.

While many of the stories and ideas were collected over 37 years as a military spouse, I enlisted the help of many senior Air and Space Force spouses to provide their perspectives and to share stories of the greatest challenges they faced. If anything, I hope their stories encourage you in challenging times, spark creativity and joy, and even bring a few laughs as well.

As you read through these chapters and digest these stories, ask yourself: what are the expectations I place on myself? What do I think others expect of me? How will I contribute to the command team and build up the unit? What kind of example do I want to set for others in the unit? How much can I engage and when should I ask for help from other spouses? It is all a balancing act, and it starts with an honest look at yourself and the command team.

Enjoy the journey!

MRS. DAWN GOLDFEIN
Spouse, USAF 21st Chief of Staff

Acknowledgments

First and foremost, I'd like to thank Dr. Paul J. Springer, who is a co-author for this project. I am grateful for his efforts in starting this book with my husband and for laying the groundwork to make this book possible. Through all of his efforts to research, interview, and collaborate with our military spouses, we now have a written work that brings a whole new perspective to command. Believe it or not, his homemade bacon is almost as good as his writing!

I especially want to thank my right-hand woman, co-author Capt Katelynne "Starbuck" Baier, who worked countless hours to interview spouses, contribute content, and provide editorial oversight to this book, all during a worldwide pandemic.

I also need to recognize the following senior Air and Space Forces military spouses for their perspectives and amazing insights shared in this book:

> **Mrs. Mollie Raymond**, Spouse of the 1st Chief of Space Operations, Space Force
> **Mrs. Nancy Wilson**, Spouse of the Vice Chief of Staff of the Air Force
> **Mrs. Tonya Wright**, Spouse of the 18th Chief Master Sergeant (CMSgt) of the Air Force
> **Mrs. Leah Thomas**, Spouse of the Pacific Air Forces (PACAF) Deputy Commander
> **Mrs. Stephanie Johnson**, Spouse of the PACAF Command CMSgt
> **Col Kristen Thompson**, dual-military active duty spouse and current Military Assistant to the Under Secretary of Defense for Policy
> **Mr. Alan Frosch**, Spouse of the Air Mobility Command Deputy Commander
> **Mr. Eddy Mentzer**, Associate Director of Children, Youth, and Family, Office of the Secretary of Defense (OSD)
> **Ms. Michelle Padgett**, Director for Family Wellness, Joint Artificial Intelligence Center, OSD

Next, Mr. Jason and Mrs. Jodi Womack inspired me with their book, *Get Momentum: How to Start When You're Stuck* and the seminars they teach as a part of the Wing Commander and Group Commander Spouse course at Maxwell, Air Force Base. Thank you for all you do to raise up our new command and lead spouses!

Finally, I want to thank the love of my life, Gen Dave Goldfein. Your big heart for our Airmen and their families shines through everything you do, and I have loved every moment of "sharing the journey" as a military spouse by your side.

Abstract

A unit's command team is the partnership among the commander, the senior noncommissioned officer (NCO), and a volunteer lead spouse. As the primary advisor, ambassador, and advocate for the spouses and families of members in the unit, finding the right person to undertake the important role of volunteer lead spouse is one of the most important decisions a commander will make. Once a spouse in the unit decides to take on the role, it can be challenging and incredibly rewarding to navigate working with military leadership, state or local government, base programs and organizations, and other military spouses to take care of families. This book captures "words of wisdom" collected by Mrs. Dawn Goldfein, spouse of the 21st Chief of Staff of the Air Force and Gen David L. Goldfein over their 37-year career. For command teams that seek to understand and leverage the military "spouse network" of command, lead, key, and key spouse mentors within their unit or their installation, it offers a treasure trove of useful ideas and stories.

Chapter 1

The Volunteer Lead Spouse
Qualities, Roles, and Definitions

To handle yourself, use your head; to handle others, use your heart.

—Eleanor Roosevelt

A unit's command team is the partnership among the commander, the senior NCO, and the volunteer lead spouse. As the lead spouse is the primary advisor, ambassador, and advocate for the spouses and families of members in the unit, finding the right person to undertake the often-thankless task of the lead spouse is one of the most critical decisions a commander will make. Every unit has unique stressors that drive varying needs of the Airmen and their families, resulting in different demands for the lead spouse. Still, there are several essential qualities lead spouses must possess to be most effective in their role.

Lead spouses are people of integrity who speak truth to power. Qualities like empathy, a deep well of compassion, and the ability to listen well ultimately make them a trustworthy focal point for other spouses or significant others. Intuition enables them to pick up on things that are unsaid within the unit. Finally, combining these traits with effective communication skills will ensure that the commander and senior NCO clearly understand problems and how the command team can help. For example, for military members whose spouse's full-time employment is a significant financial concern for their family, commanders may be unaware of restrictions and barriers to spouse employment and what tools they must bring to bear against these issues. In instances like these, members of the unit community might be far more comfortable speaking to the lead spouse than to the commander about problems they are encountering.

The lead spouse often acts as a clearinghouse of information and can provide a "civilian conscience" for the commander. They possess exceptional judgment—sometimes, the things that they choose not to raise with the commander are as crucial as those that they do verbalize. Discretion in handling sensitive spouse or family issues is one of the toughest yet most rewarding aspects of being a lead spouse. Knowing how to leverage unit and base resources to assist families amid times of struggle is critical to the role of lead spouse.

The role of the lead spouse can be summarized as an advisor, advocate, and ambassador for the unit's Airmen and their families. First, as advisors, they provide wisdom and perspective on how to best support military families not only on their command team but also in local community leadership and base organizations. Second, as advocates, they provide the "military perspective" to the conversation, such as sitting on the local community school board or volunteering at a nearby nonprofit. They also advocate on behalf of spouses and children to military leadership at all levels to address policy change and resource support. Finally, as ambassadors, the entire command team represents the USAF not only to local, city, and state officials but also frequently to international partners and allies. As a spouse liaison, lead spouses can work with their counterparts to enhance relationships and create opportunities for Airmen and their families living abroad.

There is no substitute for having a lead spouse on the command team. Often, it will be the spouse of the commander, also known as the "command spouse," but that does not have to be the case. Perhaps the commander's spouse works full-time or is homeschooling children, or the commander may be unmarried. In such situations, the role will need to be filled by someone else, and the role can still be fulfilled for a successful command tour. If a commander is not married or the command spouse does not or cannot volunteer, the commander should ask the leadership in the unit for recommendations. If nobody steps up to handle the position, the entire unit will suffer, particularly in the morale department.

Admittedly, as an elementary school teacher and a mother of two daughters, I chose at times to work outside the home throughout the 37 years my husband was in the service. The decision to commit to the role of lead spouse starts with an honest conversation with your partner about the level of involvement you both desire, including setting goals and priorities at the outset of a command tour. Managing expectations is vital to successful command assignments. If you are a command spouse and choose not to be the lead spouse, it is perfectly fine! Instead, commit to helping your partner find an individual who can fulfill the role.

One of the most common sources of confusion when it comes to the "spouse network" is the differences in roles between command spouse, lead spouse, key spouse, and key spouse mentors. Each has a vital part to play in fostering community in a complementary way, but again, there are not any hard and fast rules for spouses who take on these roles. Before fully getting the conversation started regarding volunteer military spouse service, it is essential to understand the following functions and definitions.[1]

Command Spouse

If a commander is married, his or her spouse is a command spouse. Again, this spouse might be working full-time, dealing with health or family issues, or homeschooling their children—there are any number of reasons that do not obligate them to take on the role of lead spouse. That being said, the command spouse ought to help select a lead spouse and, once selected, be intentional about supporting and deferring to that lead spouse. Be careful not to try and take back what you have given up! Be honest with yourself with what you can handle, and be consistent in helping support those who step up to volunteer.

Lead Spouse

This spouse acts as the unit-level advocate, ambassador, and advisor on behalf of unit spouses and their families. The lead spouse is typically married to an officer in a senior position in the unit. He or she is the primary lead for unit family morale and actively supports unit spouses with family issues. The lead spouse can also become a designated key spouse mentor.

Key Spouse

This role is a mandatory, commander-appointed volunteer unit position. Key spouses attend the official training at the Airman and Family Readiness Center (A&FRC), create and maintain a program in accordance with Air Force Instruction 36-3009, *Airman and Family Readiness Centers*, and are trained to respond and assist with family issues in response to deployments, crises, and other hardships families may face.[2] Although it is up to each commander, it is best if different spouses fill the lead spouse and key spouse roles. An example from the squadron level is that a lead spouse might be married to the squadron's director of operations (a major or lieutenant colonel). In contrast, the key spouse might be married to one of the unit NCOs (a staff sergeant), which results in a diversity of thought and representation of rank across the unit.

Key Spouse Mentor

This position is a commander-appointed, volunteer position. Key spouse mentors complete the official mentor training at the A&FRC and act as the primary coordinator and advocator for unit key spouse initiatives and needs. An example of how a key spouse mentor can help is if the wing or delta lead spouse acts as the key spouse mentor for all subordinate unit key spouses.[3] They can advocate for key spouse issues directly to the wing or delta commander. It is common for lead spouses to double as key spouse mentors to streamline this process.

To illustrate how vital the role of the volunteer lead spouse is to a unit, read this tale from senior Air Force spouse Mr. Alan Frosch who took on the role of the lead spouse when his wife assumed command in the immediate aftermath of the 9/11 attacks. He explains:

> My wife had just taken command of a tanker squadron in the continental United States after 9/11. She rapidly deployed forward with most of the squadron, and suddenly I was in charge of the morale and supporting the spouses and families left behind. Being a male spouse had its complications, adding friction to necessary, open, and transparent communications with the families. I felt it was important that I not make face-to-face contact alone with female spouses, so each time I went to someone's house, I would have my operation officer's spouse meet me in her car in front of the home, and we would walk in together then leave in our cars. That said, we made a good team!
>
> I created a squadron spouse roster and sent it to everyone, so everyone would have a list of other spouses that were going through the same things and to help everyone make new friends. Every week, the operations officer's spouse, and I would call everyone on the roster. I started with the "A's" one week, and I would be lucky to get to the middle of the alphabet after spending more than an hour on some phone calls. The operations officer's spouse would start at the "Z's" and would work her way to the middle. We would change the next week so that everyone would have our ear at least once a week.
>
> Some spouses worked outside the home, and they would put in all the overtime they could to keep busy. Some stay-at-home parents would stay home with their kids, watching the news all day to see if they could get closer to their spouses that way. It wasn't easy to bring everyone together as a support group.
>
> Previous to our arrival, the officers' spouses met once a month for a coffee. I changed the coffee into a weekly dinner hosted by volunteers in their homes, with a lunch in the middle of the week, and a Saturday at the chapel annex with everyone invited. Of course, I expanded the invitations to make certain that enlisted families were included as well—we were all in the same situation together.
>
> The working spouses said they couldn't take off work to attend anything, but when asked where they went to lunch, they opened up and invited all the available spouses and children to attend. We found a lot of new places to eat that we had never heard of before!
>
> The chapel annex provided a large kitchen and a new big-screen TV with a DVD player! That enabled us to have the older children watch the younger children and keep them busy with the DVDs that some of the families brought in, while we cooked in the kitchen. I "let loose" one of our spouses who was a culinary arts major to run these, and we all learned new recipes.
>
> I took a lot of photos and developed a website for the squadron to share info and pictures. I posted new pictures weekly for my spouse to download at the deployed location, where she would print the photos and place them under the briefing tables.

Our Airmen knew why they were deployed, and that their families were well taken care of. Despite the hardships and loneliness, it turned out to be a good, growing experience for all, in my opinion. Spouses that would have normally not gotten to know each other became best friends and even started naming babies after each other afterward![4]

The story that Mr. Frosch relates demonstrates one of the main aspects of the lead spouse role—making sure that families have an opportunity to raise issues before they become problems and creating social outlets for family members as a means to maintain morale.

Ultimately, the division of labor and the number of spouses in these roles may change significantly from unit to unit or even base to base; each command team and unit network will develop their dynamic tailored to the needs of every unit's circumstances. Get the balance right, and it can have an enormous effect on the morale of the entire organization. As a command team, ask yourself: *Have you given serious thought to your volunteer lead spouse, and what role or responsibilities will they take on?*

Notes

1. Air Force Personnel Center, "Key Spouse Program," https://www.afpc.af.mil/.
2. Air Force Instruction (AFI) 36-3009 *Airman and Family Readiness Centers*, 29 May 2020.
3. The US Space Force states that a delta "will have three field commands and subordinate units designated 'deltas' and 'squadrons,' . . . [and] deltas will be organized around a specific function such as operations, installation support, or training. Squadrons will be created within the deltas to focus on specific tactics." For more information, see Marcia Smith, "Space Force Unveils Organizational Structure," SpacePolicyOnline.Com, 30 June 2020, https://spacepolicyonline.com/.
4. Mr. Alan Frosch, interview by Dr. Paul J. Springer, 31 October 2019.

Chapter 2

Words of Wisdom for the New Lead Spouse

Success is not what you have, but what you are.

—Dr. Bo Bennett

Leadership is a humbling privilege—one I reflect on often when I think of our experiences over the last 37 years of service. Whenever I was able to fulfill the role of the lead spouse or ever got the chance to speak to new lead spouses, these were some common "words of wisdom" I always kept in mind:

Stay positive, even during tough times.
- Negative talk, gossip, and complaining spread like wildfire in organizations, especially if they come from the lead spouse.
- Always try to remain positive when passing on comments. When you have a bad day, find and connect with those family or friends outside of the Air Force world.

Be approachable.
- Command spouses can be perceived as a "big deal" by younger or more junior spouses. As the command or lead spouse, it is very likely you will have to initiate conversations and get the socialization rolling. You will need to intentionally demonstrate you care about what the other spouses have to say.
- It is a cliché, but it is always worth saying: people will not remember what you say, they will not remember what you wear, . . . but they will remember how you make them *feel*.

Be appreciative.
- Many people on the base and in your partner's unit will bend over backward to help you and your spouse. Acknowledging their efforts with a genuine sense of gratitude will go a long way.

Be yourself.
- Spouses do not have a rank. People immediately see through those who try to use their partner's position. We have all seen spouses who "wear their partner's rank," and it never goes over well. People can see through a phony persona, and they will certainly never confuse a spouse with a ranking military member.

Be present.
- Never underestimate the power of your presence at unit functions, spouse gatherings, or base-level events.
- Often, just showing up can speak volumes. It demonstrates that leadership cares and supports the efforts of those involved.

Be aware of your surroundings.
- One consequence of being a lead spouse is that you must be cautious about casual statements—they can be taken as the ground truth and magnified entirely beyond your control.
- It is always good to keep a lighthearted attitude in command. However, sometimes laughing and joking can be problematic when third-party observers are not part of the joke or when it is a solemn event.

You are always "on," and people are always watching.
- When I chose to be the lead spouse during my husband's command tours, it was vital for me to set a good example, not only because I represented my husband and the unit, but also I was representing the Air Force.
- At the group and wing or delta level, there is more involvement with local community leaders or international partners, in addition to the men and women on the base. Lead spouses can represent the Air Force well by attending official functions, entertaining staff, socializing with subordinate commanders and spouses, leading spouse activities, and hosting events at their house, and so forth.

Be realistic.
- If you choose to be a volunteer lead spouse, that is great—the Air Force ***absolutely*** needs people willing to offer up their time and their efforts to care for fellow members.
- Commanders cannot expect that their spouses or senior NCO spouses will be willing or able to make such a sacrifice to serve if they have employment or family commitments.

Life is always a balancing act, but being at the command level adds another element to the mix. In a perfect world, life should break down into balanced areas: 25 percent of your time is spent on you and your partner, 25 percent is spent on yourself, 25 percent is spent on family, and the other 25 percent is spent on work. But when in command, life never seems to work out quite that way, and when push comes to shove, one of those categories is going to suffer. For example, I was able to stick to and maintain self-care during our time in command—getting to the gym, eating right, and keeping mentally active.

However, we struggled in command when it came to activities as a couple. A typical date night included attending a unit dining-in or Airman Leadership School graduation.[1] We both knew that command was not forever, and we took the necessary steps to handle the temporary changes in our dynamic.

The real challenge was the family time issue—while our rule was always "family first," it was just not always feasible. We still found a way to take leave and have a family vacation during command tours—it was difficult sometimes, but it was also necessary. You want to end your command with the same family you started it with, and that is not always possible if you do not take care of each other at home. Members of the military do not set aside their personal lives to "don the uniform," and we should not expect them to do so. As commanders or lead spouses, families serve as our most essential support network. Unfortunately, it can be easy to take them for granted, especially during particularly stressful periods of work. It is so easy to allow the "urgent" to crowd out the "important." If we all set realistic expectations for ourselves and our team members, we will be more efficient and more effective as a result.

There was a period when my husband was a squadron commander where the unit was on "high alert" waiting for orders to deploy any day. The squadron was training day and night, and this meant my children, and I hardly ever saw my husband awake at home.

After several weeks of this, I saw how much my girls missed their father and decided it was time to do something about it. I called my husband to warn him we were coming over to his office at the unit, and we packed up toys and a picnic basket and drove over. For one hour, we spent time as a family eating, drawing with markers on the whiteboards, and playing with Barbies. The time still felt too short, but it was much needed.

At my husband's farewell ceremony, we discovered a Barbie had been left under his desk and was discovered by some of the unit's lieutenants. Needless to say, that Barbie did not look the same when we got it back!

That assignment taught our family that sometimes the needs of the unit win out over family time, but there are little things you can do to help bring back some balance to life.

While we as spouses volunteer for this military lifestyle, our military children are drafted. Every move, every new school, every time we uproot them from their friends, the challenge is to help them find the positive in each change. The challenges our military children face are amplified even more once you step into a position of leadership. The "glass house" affects everyone, but it can be especially difficult for children.

When my spouse became a wing commander, and our children attended an on-base high school, a teacher connected the dots and announced to the whole classroom, "Oh, you are the wing commander's kid!" Talk about worst nightmare; the greatest desire of any child at that time is just to fit in. There were giant crocodile tears that night when they came home from school, sobbing and saying, "Why can't you have a ***normal*** job?" There have been many times we felt like the worst parents ever.

Now, many years later, both of our children separately voiced that they are grateful for the lessons learned and skills gained as military children. First, they realized they could make anywhere they live a home; second, they can make friends wherever they go; and finally, they know how to recognize when others are struggling and include them.

As a new command team, be extra thoughtful about how your new roles may affect your children. Have an honest conversation with them about what is going to change and how involved they want to be, and help them set expectations.

One final thought for our new lead spouses and command teams is what we like to call the "news hour test." In many cases, your judge advocate general or legal counsel can be your best friend. If there is ever any doubt or gray area about an event or available resources, ask yourself, "Would this make us end up in the nightly news? What are the ways this could be viewed negatively?" We would always hope that everyone has the best of intentions, but sometimes safe is better than sorry.

As an example, my spouse and I received an invitation from one of our civic leaders who was hosting a barbeque party. We previously attended several events with them and genuinely enjoyed spending time together. So when we got the invitation, I jumped at the chance to go. My spouse reached out to get a few more details about the gathering; as it turned out, it was a political fundraiser for a local congressman. Our attendance could have been construed as advocating for one political candidate over another that would represent the district in which the base resided. Thank goodness, my spouse double-checked!

Keep in mind; there is a difference between legal advice and following core principles and instincts as a command team. Never be afraid to do what you know to be right for your Airmen and for the culture you are creating in the unit. Legal is there to advise, but they do not command; that authority and responsibility will always rest with the commander. Thankfully, if you have built the right command team, you can work together to navigate the gray areas and accomplish wonders.

Notes

1. A dining-in is "a formal dinner for the members of a wing, unit, or organization. Although a dining-in is traditionally a unit function, attendance by other smaller units may be appropriate." For more information, see Air Force Pamphlet (AFPAM) 34-1202, *Guide to Protocol*, 8 May 2019, 44–57, https://static.e-publishing.af.mil/.

Chapter 3

Transitions: How to Step In and Out of Command

Not all of us can do great things. But we can do small things with great love.

—Mother Teresa

Where to Start: Stepping into Command

If you are a command spouse who chooses to take on the role of lead spouse, start by having a heart-to-heart conversation with your spouse on the level of involvement you wish to commit. To help you both think through your priorities and goals, here is an exercise developed by Mr. Jason Womack, author of *Get Momentum*, on how to start:

1. Take a large piece of paper and draw out eight boxes, with four boxes on top and four boxes on the bottom. Label each box as a quarter (three months) of the two-year command, with the first year on top and the second year on the bottom.
2. In each box, mark down significant family events. Will you be finishing a degree? Children graduating? Significant family travel plans? Map it all out.
3. Next, think about the goals you would like to accomplish for yourselves, for the family, and the unit by the end of the two-year assignment. These can always be adjusted as you settle into the job, but you need a place to start.
4. Finally, break those goals down quarter by quarter until you can see the whole picture.[1]

This process is simple yet effective. It provides a robust visual platform to discuss priorities and life balance pragmatically.

If you are a lead spouse not married to the commander, the exercise above can be tailored to use when discussing unit goals and priorities. Initiate a meeting with your commander and senior NCO to discuss your role and define expectations. What are the commander's priorities for the unit as a whole? Most commanders usually have these written down in a guidance memorandum. Find significant goals that apply to the quality of life as well as family issues and use them to launch a discussion with the commander. Highlight agreed-upon focus areas and develop a timeline to accomplish the commander's goals. Also, develop criteria you could use for problems or ideas that you would like to bring to their attention.

Although the discussion about priorities and expectation management can start well before you arrive at the new base, once you are in place, it is time to approach the unit and the whole base with an open mind. Be slow to make a change when you first arrive; it is tempting to dive right in and execute your plan. However, you may discover something new or unexpected that changes your priorities. The easiest way to make a smooth transition is to observe. A simple way to do so is to request a tour of the base. Every base has a unique story to be told with many historic buildings, missions, and upcoming changes that may affect your unit.

When my spouse became the wing commander at Holloman AFB, we requested a tour of the base within the first two weeks of arriving. I remember when the bus turned toward the flight line, and I found myself rolling my eyes. "Oh no," I thought, "another flight line tour." However, the bus drove on and on past any runways or airplanes or hangers into what looked like an open desert. Off in the distance, I could see something very shiny, but could not make out what it was. As we approached, it came into view: the shiny metal was from giant metal cages! I was introduced to the Alamogordo Primate Facility, or better known as the "monkey farm" on base. These chimpanzees included those that flew into space during early exploration and testing missions. After many years of service to the nation, these chimps were given a retirement home at Alamogordo, with every service from dentists to playdates. I fell in love with the "monkey farm," and it quickly became a favorite stop on every base tour we give to visitors.

Part of touring the base is connecting with experts at the A&FRC, the chapel, mental health, medical facilities, and even the dorm managers. It may feel like a firehose; do not stress about remembering every detail! It is enough to put a face to a name, get signed up for email distribution lists or Facebook pages, and ask about available opportunities for your unit.

Keep in mind that you can leverage your unit key spouse, who is specially trained to connect spouses to resources both on-or off-base. Your goal as the lead spouse is to find ways you can plug-in to the base to advocate for your unit's spouses, families, and loved ones. This advocacy requires a proactive approach; do not wait for others to come to you.

Getting to know your "spouse network" and wing or delta leadership team is a significant next step, and it can happen at the same time you are becoming familiar with your new base. Meeting with lead spouses, command spouses, key spouses, and key spouse mentors from across the base could be the fastest "spin up" you will ever receive. If it is your first time at an event hosted by a wing or delta lead spouse, they may officially welcome you to the group, so be prepared with a short introduction and thank you "elevator speech."

Do not forget to love your predecessor; it is human nature for people to compare you to the previous command team and to want to endear themselves to you, and sometimes this results in them bad-mouthing those who came before. Although it may feel good, or it may seem like it gives you an "honest" perspective about the unit, you do not know all the facts. Just keep in mind that when it comes to transitioning out, the same things might be said about you. Try to squash a comparison mindset in the unit leadership and make it about "moving the ball forward."

As you connect with your base leadership and lead spouse, try to find out how your unit fits into the mission of the base and the numbered Air Force or major command for the Air Force or deltas, garrisons, and field commands for the Space Force.[2] Organizational charts and mission statements are your friends! As Simon Sinek famously argued, sometimes the best place to start is with the "why" behind the long workdays, short (or shortened) weekends, and frequent travel for the military members in your unit.[3] Understanding how your unit fits into the big picture will prepare you to help other spouses and significant others get through challenges and struggles together. In a way, this makes you a translator for spouses who may be new to military life and are intimidated or confused by military jargon. Do not pressure yourself to become a military expert, but be prepared just enough to answer the "why."[4]

Finally, as an ambassador for the USAF, reach out and meet with your community's civic leaders and their spouses—they will be among the most significant resources for the unit and the base. These civic leaders may also invite you to participate in local community nonprofit fundraisers or events. Keep your goals and priorities in mind as new opportunities arise, but do not be afraid to try something outside your comfort zone!

When we arrived at our new base, I was invited to participate as a member of the local school board. I'd never sat on a school board before, but as an early childhood educator, it was something I was passionate about. It was important to the community outside the base to have a military perspective since there were so many military children who attended off-base schools. One of the first issues the school board tackled was changing the course structure called a "block schedule" at the high schools. This block schedule made it so only certain classes required for graduation were available during the fall and others during the spring. This was a nightmare for military children showing up at any month of the year, and it prevented them from graduating on time. With the help of many local community members as well as military advocates, the school system expanded their course schedules and created more flexibility in the graduation requirements to address military child education needs better.

Setting Your Successor up for Success

As the outgoing lead spouse, you can help set up the next spouse for success. One of the most effective ways to do this is by updating or creating a continuity book for the future lead spouse, including:[5]

- Latest information about the installation and upcoming projects
- Latest changes and updates to quality of life and resilience programs on-base or online
- An annual timeline with the regular or reoccurring base or wing/delta events
- A spouse directory
- Any relevant unit or base social media accounts
- A current organizational chart
- A list of civic leaders and spouses as well as their bios
- Your favorite local dentists, veterinarians, hairdressers, barbers, etc.

As a command team, when you reach your final one to two quarters in command, this is the time to take an honest look at the initiatives, projects, and processes you have instituted. At that point, you are carrying a tool belt with two spray bottles: one with "weed killer" and one with "miracle grow." What does this mean? If something worked well, make sure you pour even more resources and effort into that project to make sure it continues to take root in the organization. If something did not work well, it might be time to get rid of it so that your successor has one fewer issue on his or her plate. Whatever you decide, document it, and put it in your continuity book for the next team.

When it is appropriate, reach out to the incoming spouse early to establish communication. Share the good and the bad with the new lead spouse. Be honest with him or her, but also try to filter your advice to only what he or she *needs* to know. It is easy to try to describe "how" to do things, but highlight "what" needs to be done and let your replacement spouse figure out the "how." He or she might have a different leadership style and goals from you; that is the nature of changeover. Treat your replacement the way you wish you had been treated. Also, remember that when the time comes for the change of command ceremony, it is about the new command team.[6] When participating in a change of command, make sure to not "loiter" or stay involved in the unit and prepare your organization to shift gears to the new team.

When my spouse and I prepared to leave squadron command behind, we were anxious to adhere to the adage of not "hanging around" after the change of command ceremony. In fact, we overprepared: we backed our car to the exit stairs behind the stage where the outdoor change of command ceremony was taking place and planned to hop in and drive away with our two daughters the moment the ceremony ended. Well, the ceremony concluded, and we walked together off the stage into the car and started to drive away. I happened to look up and glance in the car mirrors, and I shouted to my spouse, who was driving, "Oh no, the girls!" We were in such a hurry; we forgot to check and see if our daughters made it in the car. Instead, they were sprinting behind the car, trying to catch up, yelling, and waving. We all laughed once they caught up and jumped into the car, but that experience gave a whole new meaning to "getting out of dodge!"

Notes

1. Mr. Jason and Mrs. Jodi Womack lead a practical yet incredibly insightful exercise as a part of the seminars they teach at the Wing and Group Commander Spouse Course at Maxwell AFB. For more information, see *Get Momentum: How to Start When You're Stuck* (Hoboken, NJ: John Wiley & Sons, 2016).

2. The Space Force's new field structure "effectively organizes space forces to fight in place within mission deltas and aligns installation support functions within garrisons. Air Force expertise, units and personnel will execute installation support functions under the command of the O-6 garrison commander through Air Force mission support groups, medical groups, and special staff." For more information, see Space Force Public Affairs, "Space Force Begins Transition into Field Organizational Structure," 24 July 2020, https://www.spaceforce.mil/.

3. Simon Sinek, *Start with Why: How Great Leaders Inspire Everyone to Take Action* (New York: Penguin, 2009).

4. Sinek, *Start with Why*.

5. A continuity book is "a reference document produced by an individual to share relevant information concerning a duty or position on which he/she has knowledge. It is normally produced for an individual assigned to take over that duty or position, such as a replacement NCO [or spouse] designated to substitute a departing squad leader. If a soldier has more than one duty, he/she should have several continuity books." For more information, see "Use the U.S. Army's Approach to Continuity Planning," TechRepublic, 27 September 2006, https://www.techrepublic.com/.

6. A change of command is "a ceremony that allows subordinates to witness the formality of command [leadership] change from one officer to another. The ceremony should be official, formal, brief, and conducted with great dignity." For more information, see Air Force Pamphlet (AFPAM) 34-1202, *Guide to Protocol*, 8 May 2019, 22–24, https://static.e-publishing.af.mil/.

Chapter 4

Leveraging the Spouse Network

Magic happens when you connect people.

—Susan MacPherson

So you have a plan and are ready to dive in as the volunteer lead spouse! What now? Throughout years of base visits, international travel, and videoconferencing sessions, I have found some recurring, successful programs and processes that lead spouses help establish and maintain. Some aspects that program units have in common are family sponsorship programs, squadron and community events, fundraisers, spouse mentorship and spouse clubs, and community outreach. In the midst of it all, there is also coordination between the lead spouse and the key spouse. We tried to consolidate words of wisdom for these programs and processes, but do not take these as requirements or mandates for your unit.

When military members first arrive for a new assignment, most, if not all, go through some in-processing or welcome meeting and checklist. However, before they even depart their previous assignment, it is standard practice for a unit to connect them to someone already at the unit to act as a "sponsor." This person will answer questions and assist them with settling into the area. During this time, it is critical to make contact with the spouses or partners of incoming military members (if they have one) to connect them to the local spouse network. This connection is especially crucial if the military member and their partner are new to military life. There are so many confusing and frustrating aspects about moving, in-processing, and settling into a new area, so the sponsorship program is truly essential when it comes to developing a culture of belonging, inclusion, and family within the unit. One of the first questions you could ask as a lead spouse is to inquire about the health of this program. Survey some of the families that recently arrived and find out how their move went. A common challenge for spouses and significant others is connecting to other unit spouses when they first arrive.

One useful program new military spouses should take advantage of is called Heartlink and is hosted by the local A&FRC. It is a program that covers the mission of the base, provides an introduction to the Air Force family, and offers a toolkit for preparedness. The class also provides tips for communicating within the Air Force, addresses how families can obtain services to stay healthy, and stresses the vital importance of the Air Force family.[1]

Often, the command team, including the lead spouse, needs to reach out and rely on the military member to reach their spouse or partner. This outreach can be inconsistent at best, so developing a welcome process that specifically reaches inbound spouses and partners is a great place to start.

One of the best parts of being a lead spouse is the chance to mentor other spouses. Find those young military spouses, teach them what you know, and mentor them to walk in your shoes someday. You'll earn "bonus points" if those spouses you chose to mentor do not look or sound like you but are from different backgrounds, ethnicities, or genders. If you are a group or wing or delta lead spouse, consider pursuing the selection and training to become a key spouse mentor. Additionally, there are some non Air Force specific organizations, such as the Military Spouse Advocacy Network that allow spouses to find other military spouses to mentor and guide them through major military life milestones.[2] Ultimately, mentorship is how we grow the next generation of resilient and empowered spouses as a part of our Air Force family.

Families are a critical component to maintaining morale, both as a unit and at the individual level. Military members rely upon the support of their families. Therefore, improving morale is directly impacted by improving the relationships between those families and the organization. Like so many aspects of leadership and command, keeping high morale requires daily effort. Ultimately, it comes down to trust in the command team to have the best interests of Airmen and their families at heart. When members of the unit have flagging morale, who do you think they are most likely to complain to about their problems? Sometimes it might be a fellow military member of the unit, but most of the time, the negative energy goes straight home with them, and the family is expected to act as a shock absorber. Not surprisingly, that tends to drive resentment from the family—nobody likes to see their loved ones having a rough time, especially when there's little if anything they can do about the situation. And so, the morale of individual members is soon reflected in the morale of the families. It is not just about building trust in the team; it is also about making sure families know that their military members will be able to take the time to be an active participant in the family.

Another critical component of morale comes from pride—and once again, it is directly tied to family relationships. A dual-military, active duty officer and spouse, Col Kristen Thompson, discussed the importance of unit pride in her response to questions on morale.[3] However, the same lessons that she offers make sense when viewed as a part of being a proud member of a family, too:

> You have to evoke a sense of pride in your organization, which includes pride in the mission and pride in personal excellence. Because a unit takes on the personality of its leader—your devotion to the Air Force Core Values and in motivating your unit will be

the bedrock of the unit's performance. Signs of flagging morale include a lack of cohesion, selfishness, and a lack of pride in one's work. . . . Work becomes just a job, devoid of any personal or professional satisfaction.[4]

The goal is for spouses, significant others, and family members of service personnel to be proud not only to be a part of the Air Force but also to be a part of your specific unit. That will not just magically happen as a command team. You need to provide them with reasons to be proud. At the same time, let them know that you are, in turn, proud to have them as a part of the team.

So how do you develop pride and trust among the unit spouses and partners? A crucial point of cooperation between the lead spouse and key spouse is planning and hosting morale events and community service opportunities for the entire unit. The goal is to use these events to not only bring the unit together but also to empower the unit spouses and significant others to use their talents, skills, and ideas to contribute to the team. Be willing to think outside of normal squadron activities and try something new.

This creative approach is especially needed for younger Airmen and spouses who tend to prefer virtual connections through gaming, social media, or webinars. As a lead spouse, leverage your key spouse to contact these younger spouses and significant others to learn about what issues they care about, what hobbies and activities they enjoy, and what ideas they have. Sometimes the easiest way to build morale is to find a way to serve together. Volunteering for community service projects may be a relaxed and low-pressure event to bring the unit together.

One of the most critical relationships that can also serve as a mentorship relationship is between a unit lead spouse and a key spouse. We highly encourage the lead spouse and the key spouse to be different individuals so that the unit can leverage talent and gain different perspectives. It takes a lot of work to communicate within the command team, but it also takes a lot of work to engage with spouses and handle their crises, challenges, and questions. A lead spouse can be the voice of unit spouses and families as a member of the command team. Still, a key spouse will work every day in the trenches with the families to build trust, connect them to resources, and organize events and programs to support unit morale. As a lead spouse, an important role you will play is advising your commander about who may be a great unit key spouse. Carefully consider your areas of weakness, whether it be knowledge or skills, and find someone who can fill in the gaps. Again, you'll earn bonus points if you select someone who has a different background, ethnicity, or gender than you! Diversity in representation matters and the key spouse is a unit-level role that holds a lot of influence.

When the lead spouse and key spouse relationship is effective, magical and wonderful things can happen. Col Kristen Thompson relayed a story of her partnership with an innovative and talented unit key spouse:

> I would like to share an anecdote about teamwork with my key spouse. Living in Oklahoma, there is adverse weather for much of the year, but especially during the spring when tornado season is in full swing.
>
> Back in 2017, my squadron was set to deploy right in the middle of peak adverse weather season. Knowing there would be multiple families at risk during this period, my key spouse, on her own, came up with a family care program that resembled a task force for military members and their families to use when they needed assistance. The leaders of the task force (4 stellar NCOs) separated the state of Oklahoma into four quadrants that they helped manage when a crisis hit. If a family within their zone needed assistance, that quadrant NCO (paired with the key spouse) would "deploy" or send another squadron member with expertise in that specific crisis/issue area to the family's home to help. The task force was tremendously helpful and routinely used during the deployment cycles and bad weather. It soon became the model by which all squadrons at Tinker AFB based their crisis response.
>
> The most amazing part was that this was all spearheaded by one of my squadron spouses who identified a need and then paired expertise and leadership with that need. It was incredibly innovative and well-executed. I could not have been more proud to take our task force on the road and brief all across the Air Force and specifically at the Air Combat Command (ACC) Commander's annual Wing Commander Conference. It was such an honor for our squadron to share our idea with the rest of ACC.[5]

Spouses provide a different perspective on what is vital to the success of the unit—they utilize a holistic approach and consider aspects outside the workspace. This view is also why it is helpful to have the lead spouse and key spouse positions filled by different people. As a lead spouse, you will get the chance to empower other amazing spouses to accomplish some amazing feats.

Notes

1. USAF Services Combat Support & Community Service, "Heartlink Training," https://www.usafservices.com/.
2. Military Spouse Advocacy Network, https://www.militaryspouseadvocacynetwork.org/.
3. Dual-military active duty is "when one military member marries another, the couple becomes a 'dual military' couple." See Joe Wallace, "Benefits of Dual-Military Couples," Military Benefits, https://militarybenefits.info/.
4. Col Kristen Thompson, interview by Dr. Paul Springer, 13 October 2019.
5. Thompson, interview by Springer.

Chapter 5

Common Challenges and Questions

The greater part of our happiness or misery depends on our dispositions and not our circumstances.

—Martha Washington

When you become the lead spouse, some challenges are utterly unique to the position. Over years of base visits, conferences, and command course discussions, I have received many questions from new lead spouses from squadron to wing or delta command regarding these specific issues. The following section highlights the "Top 10" recurring questions from those spouses.

To answer some of these questions, I recruited a few of my fellow senior spouses from across the Air and Space Forces to provide their inputs and stories. We will attempt to provide words of wisdom to consider, but in many cases, there is not an "easy button" or a simple right or wrong answer. I hope that these insights will encourage you to know that we have all been there and that you can and will make it through.

Question 1: How do you handle the "highs" and "lows" of command? Specifically, what do you tell yourself when things get hard?

Remember You Are Not Alone[1]

Serving at the command level can sometimes feel very lonely, but there are often others in your peer group of lead spouses or command teams that are dealing with the same stresses. Build and develop a circle of trusted friends to talk to; they will help you consider all the options available to you.

My husband received two days' notice before his first deployment. While he prepared to leave for war, I prepared to single parent two very young children. I was petrified! However, I quickly figured out that I wasn't truly alone. I was able to draw strength from the military spouses around me, and I realized our shared experiences helped us persevere.

Self-care Is Not Just a Luxury; It Is Essential[2]

When we focus on helping others, whether they are coworkers, children, members in the unit, or your spouse, it is easy to forget and neglect to care for yourself. Self-care requires a lot of discipline, but ultimately it helps you to stay strong, and it is essential on all days. Self-care can be physical—anything from exercise to a new hair cut—but it is also emotional, spiritual, and mental. Emotional care can be as simple as rereading your favorite thank-you card or encouraging notes from friends. Mental care is reading a new book or discovering a new show on Netflix. When all else fails, chocolate-covered peanuts and a glass of red wine are an excellent place to start!

Your Attitude Will Follow Your Focus

Start by counting your blessings; there is always something to be grateful for. While you should not neglect self-care, sometimes the most powerful solution is to find ways to reach out and help others. An outward focus will help you remain positive and get out of any rut you feel you may be in. Avoid displaying negativity as much as you can; it spreads faster than wildfire in a unit, especially if you are at the top. Your attitude and example will set the tone for the other spouses in the unit. Who knows? You may make a powerful, positive lasting impact on others.

As I became more involved in the military spouse community at my unit, I noticed a special quality: service-mindedness. When I embraced that "service life," things changed. My problems seemed minuscule. Military spouses who served their communities seemed to possess endless stores of strength and energy. They managed big responsibilities and multitasked in high-intensity circumstances. I saw those acts of service ended up strengthening the spouses who served as well as the communities they helped.

Once I started getting involved with other spouses or organizations that would help, I realized this institution, this thing called service is bigger than myself.

Question 2: What are some tips for effective communication with your command team? How is military communication different?

Expectation Management[3]

As discussed earlier, effective communication starts with expectation management. Initiate a meeting with your commander and senior NCO to discuss your role and define expectations. What are the commander's priorities for the unit as a whole? Most usually have these written down in some type of guidance memorandum. Find the tasks that apply to the quality of life and family issues and use them to launch a discussion with the commander on where your focus areas should be. Discuss what timeline they need to accomplish their goals, and what criteria you should use for bringing problems or ideas to their attention.

Tailor Your Approach[4]

Communication with military members tends to be direct and to the point. This method is unlike communicating with spouses, which sometimes requires a softer touch. Know your audience! When you meet with the commander and senior NCO, ask them what communication methods work best when reaching out: Email? Meeting in person regularly? Phone calls? Texting? A group chat?

Another tip is that scheduled interactions are usually much better received than pop-up meetings. Save pop-ups for when you need them (i.e., a crisis). Ultimately, it is about mutual respect for their time and yours!

Bottom Line Up Front (BLUF)[5]

Do not let military jargon be the barrier to clear communication. Before you end the conversation, clarify to ensure you understood what is meant. A great resource to help with jargon is the *Department of Defense (DOD) Dictionary of Military and Associated Terms*.[6] Also, understanding the organizational structure for your specific unit as well as where your unit fits into the chain of command for the Air Force is essential. It is especially important when it comes to advocating for resources and support for your initiatives. Possessing this knowledge will lend credibility whenever you communicate with the rest of the military leadership team and unit members.

Bring the Solution[7]

Be a filter, not a firehose! If you are meeting with a commander, that is not the time to start brainstorm solutions. Usually, if you come to commanders with a problem, they will expect you to present a solution as well. If you are having a tough time coming up with solutions or are not sure if the commander ought to know about an issue, your unit's first sergeant or senior NCO is an excellent first sounding board. The most successful lead spouses will find a way to communicate the problem and their solutions in a way that speaks directly to the commander's guidance and priorities.

Question 3: Do you handle relationships with enlisted or officer spouses differently or the same? Should there be a spouse hierarchy that mirrors the chain of command?

Spouses Do Not Have a Rank[8]

It is entirely natural for spouses to gravitate toward those who have common interests. Officer and enlisted spouses have varying experiences while in the military. Young Airmen's spouses may have unique concerns and challenges compared to those senior NCO spouses may face. Families with young children will often gravitate toward other young families. At the same time, more senior spouses may seek out those serving at similar leadership levels to gain perspective and maintain an appropriate sounding board for their struggles.

It is an entirely different matter if spouses try to leverage the rank of their military members to establish some kind of hierarchy. A good lead spouse can identify when this is happening and break down those hierarchies. Spouses who focus on rank generate resentment, create cliques, and encourage disunity among spouses and military members.

Be Approachable[9]

If you are the lead or command spouse of a unit, there will always be an initial "intimidation" factor you will need to fight. You may need to proactively reach out to younger Airmen's spouses, especially if your spouse is an officer, to make them feel comfortable about approaching you with their problems or ideas. That being said, a lead spouse cannot be the sounding board for every challenge that arises. Try to encourage spouses to solve their problems at the lowest level. This approach is not about rank; it is about empowering everyone in the unit to embrace an open, honest, and respectful culture and community.

Before there were female combat pilots in the Air Force, my husband was a flight commander in the flying squadron. The unit was preparing to deploy, and all of the unit spouses gathered at the flight line to farewell the pilots. Something immediately stood out: all of the pilots had handguns, REAL handguns, strapped to their legs. As a flight commander's spouse, I was trying to keep my cool, but it was the first time I had ever seen my husband carry a gun while flying. The danger of what he was embarking on started to set in.

The squadron commander's spouse, Nancy, saw all of our worried and somber faces, and she quipped to all of us, "Oh, they just wear that so that they can feel MANLY!" We all cracked up laughing!

This ultimately accomplished two things: first, it broke the tension and immediately placed us all at ease, and second, it made her approachable even as THE squadron commander's spouse. This turned out to be incredibly important in the coming weeks, as we all faced the challenges of deployment together as a unit.

We Are All in This Together[10]

When push comes to shove, we are all in the same boat! As military spouses, there are everyday experiences we share regardless of rank: moving, career challenges, helping children transition in schools, and so on. Our interactions need to be centered on building relationships and supporting one another. We work on projects together, we volunteer together, and we enjoy social activities together. . . the focus is on building a vibrant and caring community. Ultimately, everyone should be treated with kindness and respect.

Question 4: What do you do if you feel unit members are trying to "use" you to get to the commander? How do you handle that with grace?

Be a Good Listener[11]

This approach is not automatically a bad thing. If someone comes to you first with a problem that they want to share with the command team, it means they find you approachable and accessible, which are two excellent qualities for a lead spouse. In many cases, all anyone needs is to feel they are being heard and that someone cares about them. Repeat back to them what you understand their problem to be, and thank them for sharing the information.

Avoid Making Promises[12]

If you are a lead spouse, you likely enjoy helping people in need. This quality is an admirable trait, but be cautious about making any promises or decisions on behalf of the commander or command chief. You are not a "fixer," even if you want to help. You are a gatekeeper; do not be afraid to ask if they have first sought the help of their chain of command or used available unit resources.

Refer Them to Other Resources[13]

Unit or base chaplains, unit first sergeants, and the Military and Family Life Counselor (MFLC) at the A&FRC are all great resources to contact when it comes to additional services.

If they are younger spouses, it might be they are just unfamiliar with what organizations exist or what resources are available. Once you refer them to a resource, follow up with them and find out if it was helpful!

Avoid the "fire and forget" mentality; sometimes, those resources work the way they should, and sometimes they do not. Ultimately, it is about demonstrating genuine care and concern for that member of the unit. A military spouse details how she connected a family with valuable services:

> When we were stationed at Spangdahlem Air Base (AB) in Germany, one of my husband's Airmen, with five children, mentioned to me that his wife left him. While he worked quickly to adapt to the "new normal" of single parenting, his children did not have a support system and were struggling with the change.
>
> We connected the whole family to the local MFLC to discuss their feelings and create short and long term goals to work on together. Because we were overseas, there were MFLCs embedded within the schools, which allowed easy access for the children to the counselor.
>
> When one of the children wanted to go to Homecoming, the MFLC suggested using the "Cinderella closet," a donation-based collection of secondhand formal gowns established by the community spouses. The family adapted well to their new normal because people within the community shared connections and resources.[14]

Question 5: How do you choose what to tell a commander and what you handle within the spouse network? How do you handle interspousal problems in the unit? What do you do if spouses just don't like you?

Ask, "How Does This Affect the Unit as a Whole?"[15]

Does this issue compromise the readiness and mission of the unit? Does it meet the criteria of the commander's "red flags" or their intent for the culture of the unit? Commanders are delegated a significant amount of authority. If there is a policy or rule that causes an unnecessary struggle for families or spouses, it is absolutely in your lane to raise the issue.

Ask, "Do YOU Need to Intervene?"

Sometimes there is an issue that is between two families that needs to be resolved at their level, and in that case, maybe it is better to remain neutral and uninvolved. However, if it is a morale issue or rampant rumor that could escalate to a unit-level problem, nip it in the bud fast! Do not be too shy to grab the bull by the horns, and do not try to pretend it will just go away. Ideally, the command team is promoting a culture where relationships are built on trust, communication, respect, and honesty, and as a result, drama will not be an issue.

When I became a squadron command lead spouse, there were two spouses in the unit who obviously did not get along. My concern was how their quarrel affected unit morale as our squadron was preparing to deploy, and it was forcing different spouses and families to feel like they had to "choose sides" instead of uniting to support one another.

Although I did not know the backstory or the cause of the quarrel, I invited them both to coffee at my house but didn't tell them the other was coming. When they both arrived, I sat them down and said, "I don't know what's going on, and it's not my business, but I need your help to resolve this. I need your help to bring our unit together as a team to support our squadron as we are preparing to deploy."

I helped them focus on the bigger picture—our need to unite as families to take care of each other while the squadron deployed—and they sorted it out themselves.

Do Not Take It Personally. . . and Keep a Sense of Humor![16]

You never know what someone else is going through! Some people may be acting out of jealousy, but also you may just be an object of frustration, and it has nothing to do with you. There are times where you may feel targeted by someone, and in that instance, try to ask them directly, "It seems like this makes/made you upset. . . can you help me understand why?" A military spouse remembers how he used humor not to take things personally:

> Someone singled me out as the only male spouse in attendance at an Officer Wives Club back in the day. An older woman made a passing comment about my presence to her friends when she passed me in the buffet line, making me feel like I didn't belong there. Instead of letting it fester inside or standing my ground in anger, I opted for a witty joke in response. She laughed out loud, and that was all it took to break the ice. We quickly became friends after that![17]

Question 6: What is the toughest situation you dealt with as a lead spouse, and what did you do?

Death in the Unit[18]

Whether through suicide, operational mishap, or killed in action, a death within the unit is truly devastating. Even if you have been through a similar experience, it is essential to recognize each loss is intensely personal. Helping the family and the unit mourn is a challenging yet necessary path to navigate. I have always wished I could be stoic, a "Jackie Kennedy," and never shed a tear, but I think it is okay to cry with them. You cannot fix the situation; no one can control it—all you can do is help walk them through it.

It can be tough to know how to interact with the surviving members after a member's death. Some people do not know what to say to them and consequently try to avoid the subject altogether. As a command team, the way you address the death to the survivors is just as important as the handling of the loss itself. Do not avoid the subject; jump in and grieve with them. Sometimes the best help is taking care of the small details of life: keeping a phone log, starting a food train, providing distractions and child care to give the surviving spouse space, helping with any packing or moving, and so forth.[19] A recommended resource is *Healing Your Grieving Heart After a Military Death: 100 Practical Ideas for Families and Friends* and it is free to survivors.[20] To request a copy, email TAPS at info@taps.org.

Deployment Breakdown[21]

We had a spouse posting on social media. She was feeling overwhelmed dealing with the children while her spouse was deployed. Several spouses banded together to take turns cleaning her house, where there were some serious issues with spoiled food and dirty diapers. We also helped babysit her kids so she could spend time working with an MFLC. Unfortunately, we realized that even with the help we provided, things did not seem to improve for her. We found out she was relying on her six-year-old to change diapers and feed the younger children, and she often talked about being a lousy parent and wondered if her family would be better off without her. The first sergeant and commander both spent time with the family and decided it was in the family's best interests to bring the military member home early from his deployment. It was heartwarming to see our spouses reach out to help, but ultimately it turned into a situation that required the involvement of leadership. Our spouses tend to have huge hearts, but sometimes the best thing to do is to leave it to the experts. Be the conduit to resources and provide companionship and support throughout the process.

Divorce and Suicidal Thoughts[22]

One of the toughest situations was a mental breakdown of one of our leadership spouses that led to hospitalization and divorce. A military spouse had extreme stress because of her husband's deployment and the strain of raising the children alone. Because of these stressors, their marriage started to crumble. The weight of the whole situation led to her suicidal intentions, which resulted in her hospitalization. My heart went out to her as I tried to mentor and comfort her in the hospital, but at the same time, I was deeply concerned about her children's welfare and her deployed military member's condition. It was such a delicate situation that it felt like walking on a tightrope! There were months of attempted outreach to the family, accompanied by repeated rejections of help. We tried to counter her suicidal thoughts while also trying to protect her privacy. Other unit spouses attempted support while making sure she did not feel abandoned. It was exhausting.

The key takeaway from this situation is to have ready access to recommended resources for the circumstances, yet also recognize when you are out of your depth. I made a binder of resources and organized them by topic, picking the "Top 3" best resources so that I could respond promptly. However, not all resources are appropriate. Take the time to judge the situation and decide how quickly you need a response. Although national-level resources can be highly effective, sometimes having a local expert who can arrive in 10 minutes is more valuable.

Scrutiny as a Male Spouse

A male military spouse recounts his concerns during his wife's deployment:

After 9/11, my wife deployed with the entire squadron, leaving me behind to mentor the young spouses, officers, and enlisted. As a male spouse, I keenly felt my wife and I lived in a "glass house." I was deeply concerned about any signs of impropriety that could potentially ruin our leadership team and jeopardize her command. I was constantly analyzing what was most "appropriate" and shied away from anything that could be perceived in a negative light.

For example, I realized I shouldn't be alone with any of the spouses. I also decided that even though I was probably not going to be the shoulder to cry on, some spouses may need exactly that. With this in mind, I asked my operation's officer spouse to meet me at homes to support families, and then I would leave separately. It also meant a ***lot*** of phone calls to all the spouses.

Over time I started to find other ways to connect with the spouses. Once a week, we would meet for lunch with any available families, and every Saturday, we met with all the families in the chapel annex. We showed movies for the younger children while the mothers learned new cooking skills courtesy of a spouse studying culinary arts. . . . all the while talking together and creating new bonds of friendship leveraging our different backgrounds!

I also created a website to publish pictures almost daily of the wives and children having fun and which allowed our deployed members to download and print for their wallets and lockers.[23]

Child Neglect and Infant Death

A military spouse recounts a devastating situation for her squadron during a deployment:

While a military member in our squadron was gone on deployment, his spouse gave birth to twins and subsequently suffered from undiagnosed postpartum depression.

Their neighbors called the first sergeant after odd smells came through the shared vents of their duplex. The squadron's first sergeant asked me to join him in a welfare check to the home. When we arrived at the house, it was disgusting. The children appeared to be unclean and severely undernourished.

We immediately reported the findings to Family Advocacy, and they contacted Child Protection Services for the state. An assessment was conducted that determined there was no reason to remove the children from the home. We were told our expectations and standards for a home were "too high." This was tough to hear, but as a squadron, we still banded together to offer support to the spouse.

One week later, we received notification that due to neglect, one of the infants passed away. This was utterly devastating to hear! It was easy to second-guess our actions up to that point. . . . Should we have done more? In times like this, all you can do is to remind yourself that you did the best you could.

People will sometimes disappoint you, and you may even disappoint yourself. Use the experience to learn and grow, and then you need to move on by focusing on the next right thing to do.[24]

Question 7: What is the role of spouse clubs, and are they still relevant to today's spouse?

Spouse Clubs Exist for Social and Philanthropic Good[25]

The American tradition of spouse clubs can be traced back to Martha Washington. She often accompanied her husband and stayed in troop encampments. As the troops moved, she would find local spouses to drop in on and "have a coffee" with just to see how things were going. The term "have coffee" came from America's first military lead spouse!

Spouse clubs exist to carry on that tradition. They are all about giving back to the unit, the base, and the local community, and they exist for military spouses to support and connect. These values are timeless and still relevant today.

There Are Misconceptions About These Clubs[26]

Some common misconceptions about spouse clubs are that they are merely centers for gossip. Some incorrectly believe that they are for cliques to hang out and exclude everyone else, or for older, more "traditional" spouses to hold intimidating, fancy dinners or tea parties. The foundational values of a spouse will always be about creating a positive and supportive spouse community and volunteering in service of the unit, base, and local area. Fundraising is also a critical function of the spouse club, specifically in support of scholarships for military children in the area.

On a recent assignment, one of the local schools lobbied the officer spouse club to receive help fundraising for new whiteboards for their classrooms as their budget could not cover the cost. Thanks to their philanthropic nature, the officer spouse club fundraised above and beyond the amount they aimed for, and the entire school received thousands of dollars' worth of new whiteboards for their classrooms.

This is just one small example of all the ways spouse clubs support local community needs as well as national organizations, such as Air Force Wounded Warriors, MCEC, and TAPS, just to name a few!

It Takes Strong Leadership and a Lot of Time[27]

It takes a lot of perseverance and time to breathe life into a spouse club, especially if it is competing for the attention of some of our younger military spouses. They tend to be more independent and do not automatically identify as a "military spouse." First off, if they are not interested, that is okay! Make sure they know they always have the option to connect and join the spouse club. A military spouse describes the difficulty and the opportunity for connection:

> It can be tough to compete for the attention of younger spouses, especially when someone is so comfortable at home and not having to go out to connect with new people physically. We need to adapt and undertake specific outreach efforts for this generation.
>
> As an example, I attended a webinar that had good information for all family members on employment, mental and physical health, and resources available during the 2019–2020 COVID-19 worldwide pandemic. While going through the webinar, I realized this format

of connection and information sharing might be the new normal even after this pandemic is over. It is incredibly efficient: mentors, educators, and licensed therapists were accessible and willing to help when we needed it the most. In a virtual environment, everyone can find comfort and answers that fit their needs.[28]

Second, ask the younger spouses for help with the spouse club! They will surprise you with their creativity, resourcefulness, and capability. Finally, as the unit lead spouse, it makes a difference if you are actively involved. Running the spouse club may be a lot to handle, and delegating that role to a motivated, younger spouse may be the best option, but do not isolate yourself from the group. Be available and be visible.

Question 8: What are some tips for including younger spouses? How should I use social media to help with my role?

Ask Younger Spouses for Help[29]

If you want to encourage your younger, newer military spouses to attend unit events, put them in charge! Give them basic parameters and let them run with it. Let them make it into an activity they would enjoy. Typically, simple is better: group movie nights, pool parties, picnic dinners at a local park. It does not have to be grand to be memorable. Also, if they suggest an event that is outside your comfort zone, attend it one time and see how it goes.

Make It a Goal to Connect in Person[30]

At the beginning of their assignment, younger military spouses—especially if they are employed or new to the military—may have a difficult time getting out of their comfort zone and connecting with other military spouses. Sometimes, all it takes is just one or two military spouses who are willing to reach out in person and get to know them. Do not be offended if they do not come to squadron events, but instead, focus on creating a warm, welcoming culture. Word of mouth will work its magic! It is about connecting people, and connecting in small groups might be what works best for your unit versus the standard, large squadron events. Figure out your demographic and note which interests/hobbies/community involvement they are already connected to.

Social Media Is Essential but a Potential Risk[31]

It is crucial to keep operational security in mind.[32] Just because it is all over the news or social media in other places does not mean you ought to discuss unit or base matters on the official unit or base social media sites. It is one thing if the news media is guessing or reporting on parts of an event; it is another if you confirm it on your website.

That being said, however, social media is a powerful tool to connect with younger generations of spouses and families. Millennials (those born between 1980 and 1995) tend to interact on Facebook and Instagram.[33] Generation Z (born in 1996 and after) tend to use Instagram, Snapchat, and TikTok.[34] The medium may change, but the fundamental rules remain the same: actively share local and national resources and information about the base, bring people together via virtual/online events, and above all, STAY POSITIVE. Some might use social media as a place to vent, but negativity breeds conflict and cynicism. Avoid it at all costs. A military spouse covers the advantages of using the command's spouse Facebook page:

> Our command has a spouse Facebook page with both lead and key spouses as the administrators. We've discovered it is a **great** tool to share information, to advertise events, and to build connections.
>
> We often have a "Question of the Day," such as "What has been your favorite duty station and why?" and it gets the conversation flowing. We offer guided hikes, podcasts, mission days, etc. to help spouses learn about the local area and feel more at home.
>
> We use the group to get the word out about key spouse-coordinated events, such as picnics or holiday parties.
>
> To encourage spouses to connect to the page, we hand out candy with the name of the Facebook page in care packages for new spouses when they first arrive.[35]

Question 9: How do you address a possible situation of male spouses or minority spouses feeling ostracized in the unit spouse network? How do you encourage diversity of thought and experience while building the spouse community in your unit?

Work Toward Diverse Representation[36]

If you want all spouses to feel included and encourage diversity of thought, representation in places of leadership matters! Reach out to those spouses and ask them to share their stories, their culture, and their talents. Sometimes all it takes is offering one opportunity to someone to make a world of difference. Other efforts may take a long time and strong interaction on both sides to build trust. While you may tend toward events that you consider may be fun, if it ever comes at the expense of or blatantly ostracizes someone because of gender, background, or beliefs, it is not worth it. Strive to be inclusive in all you do. A military spouse reveals how their spouse club had a diverse group attend a function:

> You will be surprised how many events that look like they'll appeal to only a small group of people which end up drawing a large gathering. I recently heard about a trip to a local shooting range for the spouse club that ended up being their highest attended event ever. The previous spouse club events averaged around 20 participants. In contrast, this shooting club trip drew ten male spouses who never participated in any past events, as well as over 80 female spouses![37]

Keep a Sense of Humor

An appropriate amount of humor can go a long way to defuse tension while countering biased thinking or actions. More often than not, people can blunder into a situation, not knowing they are offending. Keep in mind that the ultimate goal is to build community and positive culture, not to win a fight. Try to intentionally make new members feel welcome and draw out the introverts in the group. Emphasize that the only silly question is the one not asked!

Never Assume You Understand

Diverse groups offer both the challenge and the opportunity to practice good listening. Try to hear and understand the actual problem the person is presenting. If you have created the right environment and an inclusive culture, spouses will feel appreciated for their unique characteristics and are comfortable sharing their ideas and speaking up if something is bothering them. Even then, things like language barriers can still prevent a spouse from fully connecting to a group. If this is the case, it makes a world of difference for them to connect to local ethnic groups in the area.

During one assignment, I met a spouse from South Korea who married an American Airman while he was stationed there. When they moved to the United States, she had a very difficult time adjusting with English as her second language. One of her greatest frustrations was not being able to find any local Korean-American community to connect with and help her feel at home. Her personal struggles ranged from little things like missing her hometown food to big things, such as feeling like an entire part of her personality was missing as she struggled to learn English. These feelings wore on her over time. Eventually, she found a group to connect with via social media that got her in touch with a local family. It made all the difference!

One way we helped loop in spouses with different cultural backgrounds was to start a "gourmet club," where three to four spouses would work together each month to cook recipes from their home countries. It felt like we got to travel to a new country every time we got together, and we all got to learn from each other and feel included.

Question 10: What are some resources you wish you knew existed or that you took advantage of more when you were a new lead spouse?

Knowing the right resource to turn to can be challenging! There are 50,000 nonprofits designed for military support. It is hard to know which ones are reliable and responsive, especially national-level programs. We have pooled a list of some of our go-to resources as well as directories to refer you to even more support as new situations arise. Keep in mind as a lead spouse, you can also rely on your unit key spouse to be very knowledgeable on the latest resources available to the unit for support.

The DOD, the Department of Labor, and the Department of Veteran Affairs have partnered with nonprofits, community-based organizations, and academic institutions to provide national, state, and local resources to support recovery on a variety of topics. If you ever learn of a new resource, you can submit a nomination for it to be added to National Resources Directory.

Resources

National Resource Directory
https://www.nrd.gov/
- Connecting wounded warriors, service members, veterans, their families and caregivers with those who support them
- Benefits and compensation, education and training, employment, family and caregiver support, health, homeless assistance, housing, transportation, and travel

AF Resilience
https://resilience.af.mil
https://www.facebook.com/USAFResilience/
- Spouse Resiliency Toolkit: http://spousert.wpengine.com/introduction/
- Leading Airmen in Distress: https://www.resilience.af.mil/

Fisher House
Twitter: @FisherHouseFdtn
https://www.facebook.com/FisherHouse
https://fisherhouse.org
- Allows a "home away from home" for family members to be close to a loved one during the hospitalization for an unexpected illness, disease, or injury

Tragedy Assistance Program for Survivors (TAPS)
Twitter: @TAPS4America
https://www.facebook.com/TAPSorg/
https://www.taps.org/

- 24-hour, seven days a week tragedy assistance resource for *anyone* who has suffered the loss of a military loved one, regardless of the relationship to the deceased or the circumstance of the death
- Peer-based emotional support, casework assistance, connections to community-based care, and grief and trauma resources

Spouse Education and Career Opportunities
https://www.militaryonesource.mil/education-employment/for-spouses

- DOD initiative to support military spouse careers and education
- Information on portable careers
- Education and career planning
- Resume assistance

Macho Spouse
Twitter: @MachoSpouse
https://www.facebook.com/MachoSpouse
https://malemilspouse.com/

- Interactive online community and resource guide designed by and for male military spouses
- Videos, forums, financial tips, blogs, articles, community calendar

Easter Seals
Twitter: @ESealsDCMDVA
https://www.facebook.com/ESealsDCMDVA
https://www.easterseals.com/DCMDVA/who-we-are/

- Provides education, outreach, and advocacy for military families affected by disabilities
- Helps military personnel care for their family members with disabilities and special needs

Air Force Aid Society
Twitter: @AFASHQ
https://www.facebook.com/AirForceAidSociety/
https://afas.org/

- Assists Airmen and families with financial emergencies and offers community programs that supplement childcare, educational needs, and deployment support of family members
- Administered through A&FRC on Air Force installations

Air Force Charity Ball
https://afas.org/about/events/
- The largest fundraiser for AFAS, built on the principles of Airmen helping Airmen, generating over $6 million since 2004

United Through Reading
Twitter: @UTR4Military
https://www.facebook.com/unitedthroughreading
https://unitedthroughreading.org/
- Offers deployed parents the opportunity to be video-recorded reading storybooks to their children
- Eases the stress of separation, maintains positive emotional connections, and cultivates a love of reading

Learning Counts
https://learningcounts.org/
- Demonstrates college-level learning acquired through work, volunteering, or military service
- Builds an undergraduate portfolio using the expertise gained through work to earn college credit

Military Child Education Coalition (MCEC)
Twitter: @MilitaryChild
https://www.facebook.com/MilitaryChild
http://www.militarychild.org
- Advocacy for military students in public, private, DOD, or overseas educational systems
- Peer-to-Peer mentor programs, parent, and student programs for transition assistance
- Annual training seminar, CEUs, programs for educators, parents, and students

Military OneSource
Twitter: @Military1Source
https://www.facebook.com/military.1source
https://www.militaryonesource.mil/
- Tax services, spouse employment help, webinars, and online training, MWR library information, relocation and deployment tools, and much more

National Guard Family Program
https://www.militaryonesource.mil/national-guard/national-guard-family-program
- Information on programs, benefits, resources, and access administrative documents
- Learn about family benefits, youth and community outreach initiatives, and national-level calendar events

Notes

1. Mrs. Mollie Raymond, interview, by Capt Katelynne R. Baier.
2. Mrs. Mollie Raymond and Mrs. Nancy Wilson, interviews.
3. Mrs. Mollie Raymond and Col Kristen Thomas, interviews.
4. Col Kristen Thomas and Mrs. Tonya Wright, interviews.
5. Mrs. Leah Thomas and Mr. Alan Frosch, interviews.
6. JP 1–02, *DOD Dictionary of Military and Associated Terms*, June 2020, https://www.jcs.mil.
7. Mr. Eddy Mentzer and Col Kristen Thomas, interviews.
8. Raymond and Wilson, interviews.
9. Mrs. Tonya Wright and Mrs. Stephanie Johnson, interviews.
10. Raymond and Mr. Alan Frosch, interviews.
11. Raymond, Wilson, and Wright, interviews.
12. Raymond, Wilson, and Wright, interviews.
13. Raymond, Wilson, and Wright, interviews.
14. USAF spouse, interview by Capt Katelynne R. Baier, 8 May 2020.
15. Thompson, interview.
16. Frosch and Mentzer, interviews.
17. Mr. Alan Frosch, interview by Capt Katelynne R. Baier, 3 May 2020.
18. Mentzer, interview.
19. A phone log is a document that is used to keep track of "who called, when, and (if necessary), when to call back." For more information, see Walter R. Shumm's *The Family Support (FSG) Leaders' Handbook* (Fort Belvoir, VA: US Army Research Institute for the Behavioral and Social Science, 2000), xii, https://books.google.com/. According to Megan Splawn, meal trains "are calendars that coordinate meals for our loved ones, usually when they're going through a life change." During the earlier history of the USAF, meal trains were coordinated via phone to pull together a group of individuals to provide meals to someone in need—perhaps someone recovering from an illness or who just had a baby. With today's technology, services help automate and schedule delivery. See Megan Splawn's article "Explainer: Everything You Need to Know about Meal Trains," Kitchn.com, 30 January 2019, https://www.thekitchn.com/.
20. Bonnie Carroll, *Healing Your Grieving Heart After a Military Death: 100 Practical Ideas for Family and Friends* (Chicago, IL: Companion Press, 2015), https://www.taps.org/.
21. Wilson, interview.
22. Raymond, interview.
23. Frosch, interview by Dr. Paul J. Springer, 3 May 2020.
24. Johnson, interview by Capt Katelynne R. Baier, 8 May 2020. When posting and distributing pictures of children online, please ensure that you have permission to do so and/or have a secure or password-protected unit website.
25. Raymond, Wilson, and Frosch, interviews.
26. Raymond, Wilson, and Frosch, interviews.
27. Raymond, Wilson, and Frosch, interviews.
28. Frosch, interview by Capt Katelynne R. Baier, 3 May 2020.
29. Raymond, Wilson, Frosch, and Wright, interviews.
30. Raymond, Wilson, Frosch, and Wright, interviews.
31. Raymond, Wilson, Frosch, and Wright, interviews.
32. Operations security is:

> An information-related capability that preserves friendly essential secrecy by using a process to identify, control and protect critical information and indicators. If critical information and indicators are disclosed, it could allow adversaries or potential adversaries to identify and exploit friendly vulnerabilities leading to increased risk to mission failure or the loss of life. The desired effect . . . is to influence the adversary's

behavior and actions by reducing the adversary's ability to collect and exploit critical information and indicators about friendly activities.

For more information, see AFI 10-701, *Operations Security (OPSEC)*, 9 June 2020, https://static.e-publishing.af.mil/.

33. Ashley Viens, "Visualizing Social Media Use by Generation," Visual Capitalist, 21 September 2019, https://www.visualcapitalist.com/.

34. Dennis Green, "The Most Popular Social Media Platforms with Gen Z," Business Insider, 2 July 2019, https://www.businessinsider.com/.

35. Raymond, interview by Capt Katelynne R. Baier, 17 April 2020.

36. Wright, Johnson, Frosh, and Mentzer, interviews.

37. Mentzer, interview by Capt Katelynne R. Baier, 17 April 2020.

Conclusion

Never lose sight of the fact that the most important yardstick of your success will be how you treat other people—your family, friends, coworkers, and even strangers you meet along the way.

—Barbara Bush

As a volunteer lead spouse and a command team, you are embarking on a journey that will most likely be one of the most memorable and impactful experiences you will ever have in the Air Force. You will witness firsthand the pure talent, professionalism, and skill of our incredible Airmen and families. If you succeed as a command team, you will walk away with lifelong relationships and friendships. Years will pass, and you will still receive emails or phone calls from Airmen or spouses you worked with who want to say "Hi" and update you on their progress in life and how their families are doing.

As we enter retirement after 37 years in the Air Force, I will probably have a more difficult transition than my husband. I love being an Air Force spouse and will miss our way of life, especially the people. In closing, there is a story of a lead military spouse who had a significant impact on my military life that I wanted to share.

When we arrived at our operational flying squadron, the squadron commander's spouse quickly became my personal hero. Her name was Nancy, and she was the first lead spouse who impressed upon me how vital the role is for the whole unit. She did not teach me these things by telling me, but instead, I learned them just by watching her in action.

Nancy taught me how to connect people. She organized a "phone tree" system (we didn't have cell phones back then!) to pass information to families quickly, and she mailed out a monthly flyer with the latest local events and base information.[1]

She also helped us bond together as a spouse group through simple activities, which included a weekly dinner at the local Golden Corral, where it was easy for spouses with children to come and hang out.

Nancy also showed how much she loved and cared for our spouses and families with her actions. For example, one pregnant spouse went into labor while her husband was deployed and unable to return for the birth. I got to be the videographer to document the event. Labor lasted from 3:00 p.m. to 6:00 a.m. the next morning. When we walked out of the hospital room, there leaned against the wall in a folding chair sat Nancy. She immediately jumped up when she saw us and asked, "How's our girl!?" She waited outside that room all night to make sure the spouse and the baby were both healthy.

Nancy also taught me a powerful lesson about how to help navigate a death in the unit. Upon receiving the initial notification, a pilot was missing, she brought all the spouses together. We met with the wife of the pilot, and we all prayed for good news. Nancy received a phone call with an update and requested all the spouses leave. She asked us to stay together and be there for each other. Nancy was with the wife when the command team and chaplain arrived with the bad news that her husband was killed in an accident during his flying mission. Nancy stayed with the wife and kept a phone log, organized a meal train, and accompanied the wife to every meeting in the coming weeks.

She worked hard to make sure the memorial for the pilot was precisely what the wife wanted. Nancy even stayed with the wife until the last box was loaded in the moving van. A death in a unit is the hardest challenge to face, and Nancy's actions held us all together to make it through.

Nancy continued to have a profound effect on me. Several years later, I was a squadron commander's spouse at Aviano AB. Our squadron, the "Triple Nickel" or 555th Flying Squadron, was engaged in missions over Kosovo as a part of Operation Allied Force, and my husband was gone most nights flying with his unit.

It was very early one morning I got a phone call from the wing commander, and all I remember him saying was, "It's good news, and my wife and the team will be over to your house in 10 minutes." In a whirling dervish, somehow, I got dressed, cleaned the living room, so there was not a Barbie to be found, and made coffee. . . . and all I could think about was that I was being picked up to go to a spouse's house to deliver some news about their military member. I remember thinking and praying, "Whatever it is, the person is alive, but something has happened, so they are coming to get me. God, please help me say the right thing and do the right thing. . . . make me like Nancy."

The wing commander's spouse, group commander, group commander's spouse, chief of flight medicine, and chaplain arrived at my house—all in service dress. It was only then that I realized they were there to talk to me about my husband, Dave. The group commander said, "Dave was shot down over Serbia, and he's been rescued. He is on his way here in a C-130, and we are just waiting for the call to meet the airplane. I do not know what state your husband was in or if he sustained any injuries."

After saying all this to me outside my door, supposedly I invited them all in to sit down and drink coffee, but I don't remember any of it. My mind went blank; all I could think about at that point was Dave and how I was going to break the news to my daughters.

Finally, we got the call to head to the flight line. I remember arriving with my oldest daughter and seeing a sea of "green": green flight suits and uniforms crowded the flight line as it felt like everyone at Aviano AB gathered there to welcome my husband home. The C-130 landed, the stairs dropped down the side, and out walked the most beautiful, muddy mess I had ever seen.

Why do I share that story to conclude this book? It is because at the moment when I was faced with the hardest situation I could face as a squadron lead spouse, the person I thought of—the person I reached for and wanted to emulate—was Nancy, my first lead spouse. During your current assignment, you have the chance to be a "Nancy"—or a "Ned" for our male spouses—for the spouses, partners, and families in your unit. You have the chance to inspire, lead, and mentor others. Embrace it. Enjoy it. I have no doubt you will leave this Air Force and your unit and base better than you found it.

Notes

1. According to Shumm's *The FSG Leaders' Handbook*, 11, a phone tree is "a system with levels that requires people at one level. . . to call people at the next level. . . who in turn, contact individual families."

Abbreviations

A&FRC	Airman and Family Readiness Center
AB	Air Base
ACC	Air Combat Command
AFB	Air Force Base
BLUF	bottom line up front
CMSgt	chief master sergeant
CSAF	Chief of Staff of the Air Force
DOD	Department of Defense
MCEC	Military Child Education Coalition
MFLC	Military and Family Life Counselor
NCO	noncommissioned officer
OSD	Office of the Secretary of Defense
PACAF	Pacific Air Forces
TAPS	Tragedy Assistance Program for Survivors
USAF	United States Air Force

Bibliography

Air Force Instruction (AFI) 36-3009. *Airman and Family Readiness Centers*, 29 May 2020. https://static.e-publishing.af.mil/.

———.10-701. *Operations Security (OPSEC)*, 9 June 2020. https://static.e-publishing.af.mil/.

Air Force Pamphlet (AFPAM) 34-1202. *Guide to Protocol*, 8 May 2019. https://static.e-publishing.af.mil/.

Air Force Personnel Center. "Key Spouse Program." https://www.afpc.af.mil/.

Carroll, Bonnie. *Healing Your Grieving Heart After a Military Death: 100 Practical Ideas for Family and Friends.* Chicago, IL: Companion Press, 2015. https://www.taps.org/.

Goldfein, David L. *Sharing Success—Owning Failure: Preparing to Command in the Twenty-First Century Air Force.* Maxwell Air Force Base, AL: Air University Press, 2001. https://media.defense.gov/.

Green, Dennis. "The Most Popular Social Media Platforms with Gen Z." Business Insider, 2 July 2019. https://www.businessinsider.com/.

Joint Publication (JP) 1–02. *DOD Dictionary of Military and Associated Terms*, June 2020. https://www.jcs.mil.

Military Spouse Advocacy Network. https://www.militaryspouseadvocacynetwork.org/.

Shumm, Walter. *The Family Support (FSG) Leaders' Handbook.* Fort Belvoir, VA: US Army Research Institute for the Behavioral and Social Sciences, 2000. https://books.google.com/.

Sinek, Simon. *Start with Why: How Great Leaders Inspire Everyone to Take Action.* New York: Penguin, 2009.

Smith, Marcia. "Space Force Unveils Organizational Structure." SpacePolicy Online.Com, 30 June 2020. https://spacepolicyonline.com/.

Space Force Public Affairs. "Space Force Begins Transition into Field Organizational Structure." United States Space Force, 24 July 2020. https://www.spaceforce.mil/.

Splawn, Megan. "Explainer: Everything You Need to Know about Meal Trains." Kitchn.com, 30 January 2019. https://www.thekitchn.com/.

USAF Services Combat Support & Community Service. "Heartlink Training." https://www.usafservices.com/.

"Use the U.S. Army's Approach to Continuity Planning." TechRepublic, 27 September 2006. https://www.techrepublic.com/.

Viens, Ashley. "Visualizing Social Media Use by Generation." Visual Capitalist, 21 September 2019. https://www.visualcapitalist.com/.

Wallace, Joe. "Benefits of Dual-Military Couples." Military Benefits. https://militarybenefits.info/.

Womack, Jason W., and Jodi Womack. *Get Momentum: How to Start When You're Stuck*. Hoboken, NJ: John Wiley & Sons, 2016.

Index

9/11, 4, 29

activities, 8, 9, 19, 25, 41
AF Resilience, 35
Air Base (AB), 26
Air Combat Command (ACC), 20
Air Force, 3, 4, 7, 8, 13, 17-20, 23, 24, 31, 36, 37, 41, 43
Air Force Aid Society, 36
Air Force Charity Ball, 37
Air Force Instruction (AFI), 3
Airman & Family Readiness Center (A&FRC), 3, 12, 17, 26, 36
Airmen, 1, 2, 5, 10, 18, 19, 24, 26, 35-37, 41
Alamogordo Primate Facility, 12
Ambassador, 1, 3, 13
approachable, 7, 24, 25
attitude, 8, 22
Aviano AB, 42, 43

base, 1, 2, 5, 7, 8, 10, 12-14, 17, 21, 26, 31, 33, 41, 43
breakdown, 28, 29

challenges, 9, 13, 19, 21, 24, 25
change of command, 14, 15
chaplain, 42
child neglect, 30
children, 2-4, 9-11, 13, 21, 24-26, 28-31, 37, 41
chimpanzees, 12
civilian conscience, 1
command spouse, 2, 3, 11, 24
command team, 1, 2, 5, 10, 13, 14, 18, 19, 22, 25, 27, 28, 41, 42
commander, 1-3, 9-12, 19, 20, 22-26, 28, 41, 42
communication, 1, 14, 22, 23, 27
community, 1, 2, 8, 13, 17, 19, 22, 24-26, 31-34, 36, 37
connection, 17, 31, 32
continuity book, 14
cultural backgrounds, 34
culture of belonging, 17

death, 28, 30, 36, 42

delta, 3, 8, 12-14, 18, 21
Department of Defense (DOD), 23, 35-37
Department of Defense (DOD) Dictionary of Military and Associated Terms, 23
deployment, 20, 21, 25, 28-30, 36, 37
depression, 30
dining-in, 9
diversity, 3, 20, 33
divorce, 29
dual-military, 18

Easter Seals, 36
education, 13, 35-37
enlisted, 4, 24, 29
events, 8, 10, 11, 13, 14, 17, 19, 32-34, 37, 41
expectation management, 12, 22

Facebook, 12, 33, 35-37
family/families, 1-5, 7-9, 11-12, 17-20, 22, 24, 26-31, 33-37, 41, 43
field commands, 13
fire and forget, 26
Fisher House, 35
Frosch, Alan, 4, 30

garrison, 13
Generation Z, 33
Get Momentum: How to Start When You're Stuck, 11
group, 4, 8, 12, 18, 21, 23, 32-34, 41, 42

Healing Your Grieving Heart After a Military Death: 100 Practical Ideas for Families and Friends, 28
Heartlink, 17
holistic approach, 20
Holloman AFB, 12
humor, 27, 34

infant death, 30
Instagram, 33

jargon, 13, 23
jealousy, 27
judge advocate general, 10

key spouse, 2-4, 12, 17-20, 35
key spouse mentor, 3, 18

Kosovo, 42

lead spouse, 1-5, 7, 8, 11-14, 17-21, 24, 25, 27, 28, 31, 32, 35, 41-43
leadership, 2, 7-9, 12-14, 18, 20, 23, 24, 28, 29, 31, 33
Learning Counts, 37
legal advice, 10
listen, 1
listening, 34

Macho Spouse, 36
male spouse, 4, 27, 29
marriage, 29
meal train, 42
mentor, 3, 18, 29, 37, 43
mentorship, 17-19
millennials, 33
Military Child Education Coalition (MCEC), 31, 37
military communication, 22
military leadership, 2, 23
Military OneSource, 37
military spouse, 2, 18, 22, 26, 27, 29-31, 33, 34, 36
Military Spouse Advocacy Network, 18
minority spouses, 33
miracle grow, 14
mission, 13, 17, 19, 26, 33, 42
morale, 2-5, 18, 19, 27

National Guard Family Program, 37
National Resource Directory, 35
negativity, 22, 33
news hour test, 10
noncommissioned officer (NCO), 1, 3, 8, 11, 20, 22-24
nonprofit, 2, 13

officer, 3, 4, 18, 24, 27, 29, 31
Oklahoma, 20
operational security, 32
organizational chart, 14
organizational structure, 23
outreach, 17, 18, 29, 31, 36, 37

partner, 2, 7, 8, 17, 18
partnership, 1, 20
perspective, 1, 2, 13, 20, 24

phone log, 28, 42
phone tree, 41
positive, 7, 9, 22, 31, 33, 34, 37
pride, 18, 19
promises, 25

quality of life, 11, 14, 22

rank, 3, 7, 24, 25
Raymond, Mollie, 30
relationships, 2, 18, 19, 24, 25, 27, 41
representation, 3, 20, 33
resilience, 14, 35
resources, 1, 10, 12-14, 19, 23, 25, 26, 29, 31, 33, 35-37

school, 2, 9, 10, 13, 31
self-care, 9, 21, 22
senior spouses, 21, 24
service, 2, 7, 12, 19, 22, 31, 35, 37, 42
Sinek, Simon, 13
Snapchat, 33
social media, 14, 19, 28, 32-34
Space Force, 13
Spangdahlem Air Base (AB), 26
sponsor, 17
sponsorship, 17
spouse clubs, 17, 30, 31
Spouse Education and Career Opportunities, 36
spouse network, 2, 12, 17, 26, 33
squadron, 3, 4, 9, 15, 17, 19-21, 24, 25, 27, 29, 30, 32, 41-43
stress, 12, 29, 37
stressors, 1, 29
successor, 14
suicide, 28

task force, 20
Thompson, Kristen Col, 18, 20
TikTok, 33
Tinker AFB, 20
Top 10, 21
tour, 2, 12
Tragedy Assistance Program for Survivors (TAPS), 36
transition, 12, 25, 37, 41
trust, 18, 19, 27, 33
Twitter, 35-37

unit, 1-5, 7-14, 17-29, 31-33, 35, 41-43
United States Air Force (USAF), 2, 13
United Through Reading, 37
unit leadership, 13

Veteran's Affairs, 35
volunteer lead spouse, 1, 3-5, 8, 17, 41

webinar, 31
weed killer, 14
Wilson, Nancy, 30
wing, 3, 8, 9, 12, 14, 18, 20, 21, 42
Womack, Jason, 11
words of wisdom, 7, 17, 21

young/younger spouses, 19, 29, 31, 32